FLORA OF TROPICAL EAST AFRICA

ARALIACEAE

J. R. Tennant

Trees, shrubs, lianes or (rarely) ± herbaceous, sometimes epiphytic, unarmed (at least in East Africa), with a simple or stellate indumentum, or glabrous. Leaves simple or compound; lamina coriaceous or chartaceous, often in some genera with differences in texture and outline between those of juvenile and mature parts of the plant. Stipules usually present. Inflorescences generally ample, ultimate branching very often umbellate or racemose. Flowers small, regular, often greenish-yellow, hermaphrodite, polygamous or dioecious. Calyx ± obconic, with tube adnate to ovary; free margin erect and very small. Petals 4–10, valvate, usually free, sometimes connate and calyptrate. Stamens equal in number to the petals and alternate with them, rarely more numerous, inserted (with them) on a disk. Anthers dorsifixed, dithecous, longitudinally dehiscent. Ovary inferior with 2–8 locules; styles distinct throughout or connate below into a stylopodium; ovules solitary, pendulous from the apex of each locule. Fruit a berry or a drupe, often with fleshy exocarp and an endocarp divided into distinct pyrenes or hardly distinct from the exocarp. Seeds with copious ruminate or smooth endosperm; embryo small and located near the hilum.

A widespread Old and New World family.

Many genera are cultivated as ornamental trees on account of the great beauty of their foliage. Dale (Introd. Trees Uganda Prot.: 30 (1953)) records *Didymopanax morototoni* (Aubl.) Decne. & Planch. (a native of South America & Trinidad) as introduced to Budongo in Bunyoro in 1949. The young plants were reported, however, to have been severely browsed by buck.

In the genera *Cussonia* and *Polyscias* a number of well defined pathological conditions occur. In *C. arborea* the swollen ends of spike-bearing branches and spike-rhachises seem to be particularly prone to attack by borers. Again in this species (fig. 1/10) and *C. kirkii* fruits often become pathologically enlarged reaching the dimensions of a garden pea. A rather different manifestation is evident in *P. fulva* (fig. 4/4); here a galling of the secondary branches of the inflorescence is of apparently occasional but widespread occurrence.

Note. The measurements in the following descriptions are all taken from dried specimens, and may be smaller than those taken from living plants. The family and generic descriptions may not always apply to plants from outside the area covered by this Flora.

1. Leaves more than once pinnately compound or pari- or imparipinnate; pedicels articulated below the flower except in *P. stuhlmannii* var. *inarticulata*; endosperm not ruminate . . . **2. Polyscias**
   Leaves simple and palmately lobed or digitately compound; pedicels not articulated below the flower 2
2. Endosperm ruminate; ovary with 2 locules and 2 styles (very rarely 1 or 2 supernumerary locules and styles) . . . . **1. Cussonia**
   Endosperm not ruminate; ovary with 5–8 locules and an equal number of styles . . . **3. Schefflera**

1. CUSSONIA

Thunb. in Nov. Act. Soc. Sc. Ups. 3: 210, t. 12, 13 (1780)

Trees or shrubs, generally glabrous, rarely tomentose. Leaves petiolate, simple and palmately lobed or digitately compound, often crowded, borne towards the ends of the stem or main branches; leaflets with entire to crenate margins or pinnatifid, rarely irregularly lobed. Stipules often intrapetiolar, connate with the petiole for some distance with an often once-cleft free portion. Inflorescences of spikes, racemes, umbels or panicles of umbellules; bracts subtending flowers (floral bracts) scale-like or obsolete; pedicels not articulated beneath the flower. Flowers greenish, 4–8 mm. in diameter. Calyx-margin repando-(4–)5-dentate or subentire. Petals (4–)5. Stamens (4–)5; anthers ovate. Disk flat, depressed or conical. Ovary bilocular, very rarely with 1–2 supernumerary carpels; styles 2, very rarely 3–4, connivent at base. Fruit laterally compressed or subglobose, often urceolate or subobconical to wedge-shaped (in *C. spicata*); exocarp fleshy or submembranaceous. Seeds ovoid-globose or laterally compressed; endosperm ruminate.

A genus of about 25 species confined to tropical Africa, South Africa, Madagascar and the Mascarene Islands.

NOTE. The length of fruits in the following descriptions includes the stylopodium but excludes the terminal free parts of the styles. The characters of leaves both in the key and the descriptions are taken from leaves from mature branches. The leaves of coppice and epicormic shoots and of juvenile plants are often very different, being simple and lobed instead of compound, and should not be used for identification.

1.	Flowers in racemes of umbellules, see fig. 3/1, 2 (sect. *Neocussonia* Harms)	7. *C. lukwangulensis*
	Flowers in spikes or spike-like racemes (sect. *Cussonia*)	2
2.	Flowers pedicellate, in spike-like racemes (fig. 3/5); leaves digitately compound with sessile leaflets	5. *C. zimmermannii*
	Flowers sessile or virtually so, in spikes; leaves simple and palmately lobed, or digitately compound with petiolulate or sessile leaflets	3
3.	Fertile portion of the spike abruptly distinct from the peduncle (fig. 3/3); leaves digitately compound with pinnatifid, pinnatisect or partially pinnate leaflets (fig. 2/1, 2)	1. *C. spicata*
	Fertile portion of the spike gradually merging into a sterile basal region, or spikes fertile throughout; leaves simple or if compound then leaflets crenate to entire	4
4.	Leaves digitately compound with petiolulate leaflets (fig. 2/5), glabrous or nearly so	6. *C. holstii*
	Leaves digitately compound with sessile leaflets, sometimes much attenuated basally in *C. kirkii*, or leaves simple, palmate; indumentum various	5
5.	Leaves simple, palmate, clothed with curly hairs beneath when young, becoming	

sparsely covered when mature, subglabrous or with some sparsely scattered hairs above; lamina up to 8 cm. long by 13 cm. wide, with a petiole of about 8 cm. . . . 2. *C. jatrophoïdes*

Leaves either digitately compound, or when simple and palmate up to 27 cm. long by 42 cm. wide with a petiole up to 45 cm. long 6

6. Leaves digitately compound with leaflets ($1\frac{3}{4}$–)$2\frac{1}{2}$–$5\frac{1}{2}$ times as long as wide, glabrous to pubescent . . . . 4. *C. kirkii*

Leaves simple, palmate or if digitately compound with leaflets ($1\frac{1}{5}$–)$1\frac{1}{2}$–$2\frac{1}{2}$ times as long as wide, velutinous above, tomentose beneath when young, sparsely scabrid above, curly-pubescent beneath (occasionally almost entirely glabrous) when mature . . . . . . 3. *C. arborea*

1. **C. spicata** *Thunb.* in Nov. Act. Soc. Sc. Ups. 3: 212, t. 13 (1780); Sond. in Fl. Cap. 2: 568 (1861–62); Hiern in F.T.A. 3: 32 (1877); Gilg in P.O.A. B: 342 (1895); V.E. 3(2): 782–783, fig. 326/F, G (1921); Bews & Aitken in Bot. Survey S. Africa, Mem. 5: 56 (1923); T.S.K.: 115 (1936); T.T.C.L.: 59 (1949); I.T.U., ed. 2: 34 (1952); K.T.S.: 54, fig. 10/A (1961); F.F.N.R.: 312 (1962). Type: South Africa, Cape Province, *Thunberg* (UPS, holo.)

A tree up to 17 m. or more tall, unbranched or with a sparingly branched bole*, bearing a crown of large digitately compound leaves. Petiole up to 67 cm. long and 1 cm. diameter, glabrous or minutely pubescent especially in stipular region; leaflets 6–12, sessile to petiolulate, rather coriaceous, pinnatifid to partially pinnate (fig. 2/1, 2), up to 35 cm. long by 19 cm. wide, glabrous or with a few scattered hairs; petiolules up to 8 cm. long, often narrowly winged for some distance. Spikes pedunculate (fig. 3/3); flowering region occupying one-quarter to six-sevenths of the rhachis; floral bracts up to 3 mm. long, ciliate, inconspicuous. Fruits subobconical to wedge-shaped, often constricted immediately below stylopodium, up to 12 mm. long, glabrous or minutely puberulous. Figs. 2/1, 2 (p. 9), 3/3 (p. 10).

UGANDA. Karamoja District: Mt. Moruongole, 11 Nov. 1939 (sterile), *A. S. Thomas* 3329!; Elgon, Sipi, 21 July 1924 (fr.), *Snowden* 937!

KENYA. Nakuru District: eastern Mau Forest Reserve, 8 Sept. 1949 (fr.), *Maas Geesteranus* 6181!; Machakos District: N.Chyulu Hills, 21 Apr. 1938 (fr.), *Bally* 150, 211, 428 in *C.M.* 7965!

TANGANYIKA. W.Usambara Mts., about 3 km. NE. of Bumbuli on path to Mazumbai, 10 May 1953 (fr.), *Drummond & Hemsley* 2469!; Kondoa District: southern slope of Kinyassi Mt., 6 Feb. 1928 (sterile), *B. D. Burtt* 1323!; Rungwe Mt., 10 Nov. 1910 (sterile), *Stolz* 406!

DISTR. **U**1, 3; **K**2–4, 6, 7; **T**2, 3, 5–8; Mozambique, Malawi, Zambia and Rhodesia, Comoro Is., South Africa

HAB. Upland rain-forest, upland dry evergreen forest, grouped-tree grassland; 1450–2550 m.

VARIATION. Bews & Aitken have drawings showing that the leaves from a young plant can be simple, palmately lobed, the lobes themselves having an irregularly and shallowly crenate margin.

* Bews & Aitken give detailed accounts of the conditions which govern the habit in South Africa.

2. **C. jatrophoïdes** *Hutch. & E. A. Bruce* in K.B. 1931: 273 (1931); T.T.C.L.: 59 (1949). Type: Tanganyika, Dodoma District, Chenene Hills, *B. D. Burtt* 996 (K, holo. !, EA, iso.)

A straggling shrub or sometimes a small spreading tree. Leaves closely spirally arranged at the ends of branches, or with internodes of up to 4 cm.; petiole ± as long as lamina, up to 1·5 mm. diameter, glabrous or with some hairs; lamina chartaceous, simple, palmatisect to the middle or beyond, rather small, up to 8 cm. long and 13 cm. wide; lobes 3–5, up to 4·5 cm. long and wide, broadly ovate, acuminate, with serrate margins, glabrous or slightly hairy above and with crisped hairs concentrated along the main veins beneath and at base, more evenly scattered beneath in young leaves. Flowering spikes up to 6 together, up to 20 cm. long, very lax, bearing few well-spaced flowers; floral bracts ovate, scale-like, up to 1 mm. long, glabrous or puberulous. Fruits up to 7 mm. long, glabrous. Fig. 3/4, p. 10.

Tanganyika. Dodoma District: Manyoni, Hika R. gorge on Saranda Scarp, 17 Dec. 1931 (fr.), *B. D. Burtt* 3533 !; Kilosa District: between Berega and Mlali on Mpwapwa road, 9 Dec. 1935 (fl.), *B. D. Burtt* 5428 !
Distr. **T**5, 6; not known elsewhere
Hab. Thicketed hill slopes; 1070–1360 m.

Note. It is likely that the distributional range of this species is wider than indicated by the above gatherings. Burtt states it to be " A common small spreading tree of lower *Berlinia* [i.e. presumably *Julbernardia*] zone."

3. **C. arborea** *A. Rich.*, Tent. Fl. Abyss. 1: 336, t. 56 (1847); Hiern in F.T.A. 3: 31 (1877); Harms in E. & P. Pf. 3 (8): 53 (1894); V.E. 3 (2): 782, fig. 325 (1921); Lebrun in B.J.B.B. 13: 14 (1934); T.S.K.: 115 (1936); T.T.C.L.: 59 (1949); F. P. S. 2: 356 (1952); K.T.S.: 52, fig. 10/B (1961); F.F.N.R.: 312 (1962). Type: Ethiopia, Simen [Semien], Schoata [Schuada], *Schimper* 1357 (BM, K !, isosyn.)

A tree up to 13 m. tall, with a bole up to 1 m. or more in diameter; bark thick and fissured, dark grey to reddish-grey. Leaves simple, deeply palmately lobed or digitately compound with sessile leaflets; petiole up to 45 cm. or longer and 8 mm. diameter, at first tomentose becoming tomentulose or subglabrous; leaflets or leaf-lobes 5–7, chartaceous to coriaceous, lanceolate to oblanceolate, ovate and obovate, sometimes rotund, up to 26 cm. long by 16 cm. wide, (1·2–)1·5–2·5 times as long as wide, acuminate to retuse at apex, cuneate to attenuate at base, with the margins crenate to serrulate (or subentire), occasionally irregularly so, velutinous above, tomentose below when young, becoming sparsely scabrid above and pubescent beneath (or occasionally entirely subglabrous) when mature. Flowering spikes up to about 20 together, mostly less than 10, up to 40 cm. long, sometimes galled (see note on p. 1); floral bracts narrowly lanceolate to broadly ovate, occasionally orbicular, often caudate, up to 5 mm. long, but sometimes obsolete except for tail, puberulous to densely pubescent. Fruits 3·5–4·5(–6·0) mm. long, glabrous or puberulous. Fig. 1.

Uganda. Acholi District: SE. Imatong Mts., Agoro slope, 8 Apr. 1945 (fl.), *Greenway & Hummel* 7320 !; Elgon, Bukonde, 20 Mar. 1924 (fl. & fr.), *Snowden* 850 !; Mengo District: Bugerere, Busana, Mar. 1932 (fl. & young fr.), *Eggeling* 262 !
Kenya. W. Suk District: Kapenguria, 15 May 1932 (fl.), *Napier* 1980 in *CM*. 4898 !; Uasin Gishu District: Eldoret, 13 Apr. 1951 (fl.), *G. R. Williams* 120 !; N. Kavirondo District: Kakamega, June 1933 (fr.), *Dale* in *F.D.* 3079 !
Tanganyika. Bukoba District: Karagwe, Sept.–Oct. 1935 (fr.), *Gillman* 607 !; Ufipa District: Namwele, 17 Nov. 1949 (fl. & fr.), *Bullock* 1425 !; Uluguru Mts., Mar. 1935 (fr.), *E. M. Bruce* 892 !
Distr. **U**1, 3, 4; **K**2–5, ? 6; **T**1, 2, 4–8; Ethiopia, Sudan and Congo Republics, Zambia and Rhodesia
Hab. Woodland, grouped-tree grassland; 300–2470 m.

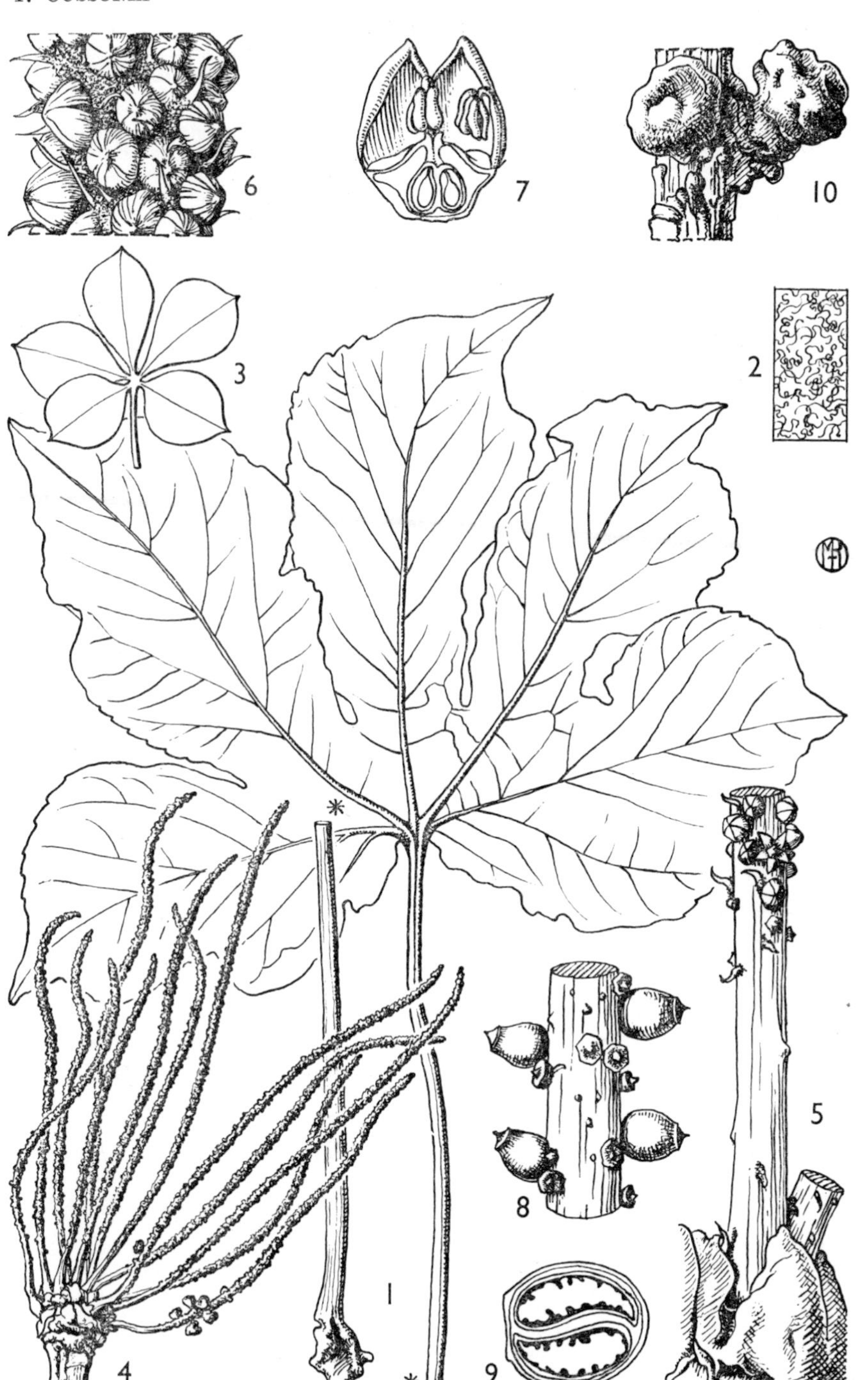

Fig. 1. *CUSSONIA ARBOREA*—**1,** young leaf, × $\frac{2}{3}$; **2,** pubescence on lower surface of same, × 10; **3,** more mature leaf (to show deeper division of lamina), × $\frac{2}{15}$; **4,** inflorescence, × $\frac{2}{9}$; **5,** base of two primary inflorescence-branches, × 2; **6,** small segment of one of primary branches of inflorescence, × 4; **7,** vertical section of flower, × 8; **8,** small portion of primary branch of inflorescence in fruit, × 2; **9,** off-median longitudinal section of fruit, × 4; **10,** galled fruits, × 2. 1, 2, 5, 6, from *Greenway & Hummel* 7320; 3, 4, 10, from *Dale* in *F.D.* 3079; 7, from *Eggeling* 505; 8, 9, from *E. M. Bruce* 892.

SYN. *C. hamata* Harms in E. & P. Pf. 3 (8): 11, fig. 3/K, & 53 (1898) & in E.J. 26: 247 (1899). Types: SE. Sudan Republic, Seriba Ghattas (near present day Wau), *Schweinfurth* 2060 (B, syn. †, K, isosyn. !) & *Schweinfurth* ser. II, 159 (B, syn. †, K, isosyn. !) & *Schweinfurth* 1891 (B, syn. †)

VARIATION. Specimens from the Sudan Republic have considerably broader, and in general longer and less coriaceous leaf-lobes and leaflets than those from East Africa. As defined here, *C. hamata* Harms is also included. A sterile specimen at Nairobi, *Busse* 2787 ! collected in 1903 in the Lindi District of Tanganyika, probably on the Rondo Plateau, has unusually coarsely crenate leaf-lobes, but should possibly be included here also.

4. **C. kirkii** *Seem.* in J.B. 4: 299 (1866); Hiern in F.T.A. 3: 32 (1877); Harms in E. & P. Pf. 3 (8): 53 (1894); Coates Palgrave, Trees Centr. Afr.: 33 (1956); Tennant in K.B. 14: 222 (1960); F.F.N.R.: 312 (1962). Type: Mozambique, Mt. Morrumbala [Moramballa], *Kirk* (K, holo. !)

A tree to 11 m. tall with a bole of up to 0·75 m. diameter; bark deeply fissured and corky. Leaves digitately compound (occasionally not quite completely so); petiole up to 87 cm. long and 9 mm. diameter, but usually much smaller, mostly glabrous or somewhat hairy in places; leaflets 5–9(–10), sessile, chartaceous to coriaceous, lanceolate to oblanceolate, ovate and obovate, up to 23 cm. long by 10·5 cm. wide, ($1\frac{3}{4}$–)$2\frac{1}{2}$–$5\frac{1}{2}$ times as long as wide, caudate to obtuse, very rarely emarginate, with a narrowly cuneate to much attenuated base, with crenate to serrate or virtually entire margins, glabrous or with scattered short hairs. Flowering spikes up to about 26 together, but mostly less than 12, up to 46 cm. long, sometimes galled (see note on p. 1); floral bracts similar to those of *C. arborea*. Fruits 4·0–5·5 (–7·0) mm. long, glabrous or puberulous.

NOTE. Simple palmately lobed leaves from juvenile parts which closely resemble the simple leaves of *C. arborea* sometimes occur. These can usually be identified by the possession of narrower lobes analogous to the narrower leaflets and except in cases of marked undevelopment, the ratios in the key may be used as a guide. *Pirozynski* 192 has a single leaf with some ± distinct leaflets but elsewhere a lamina which is merely deeply lobed. It provides an interesting transition between the simple and compound condition.

KEY TO VARIETIES

1. Flowers nearly always with 4 petals and 4 stamens . . . var. **quadripetala**
   Flowers with 5 petals and 5 stamens 2
2. Rhachis of spike with overlapping, enlarged, sterile bracts at base . . . . var. **bracteata**
   Rhachis of spike without enlarged sterile bracts . var. **kirkii**

var. **kirkii.**

Rhachis of spike without overlapping enlarged sterile bracts at base; flowers pentamerous.

TANGANYIKA. Buha District: Gombe Stream Game Reserve, Kakombe valley, 7 Jan. 1964 (fl.), *Pirozynski* 192 !; Songea District: about 25 km. W. of Songea, Nangurukuru Hill, 20 Feb. 1956 (fl. & young fr.), *Milne-Redhead & Taylor* 8838 ! & about 17 km. W. of Songea, 24 Feb. 1956 (young plants), *Milne-Redhead & Taylor* 8741 ! & Songea District, without locality, Jan. 1901 (fr.), *Busse* 807 !
DISTR. **T**4, 8; Mozambique, Malawi, Zambia and Rhodesia; also in Guinée Republic, Sierra Leone, Ivory Coast, Ghana, Nigeria and Cameroun Republic
HAB. *Brachystegia* woodland and rocky hills; 950–1060 m.

SYN. *C. barteri* Seem. in J.B. 4: 299 (1866); Hiern in F.T.A. 3: 32 (1877); F.W.T.A., ed. 2, 1: 750 (1958). Type: Nigeria, Niger R., Borgu, *Barter* 815 (K, holo. !)
For fuller synonymy see K.B. 14: 222 (1960)

NOTE. It has been reported that, in areas subject to grass fires, var. *kirkii* may exist for many years in a herbaceous juvenile state with a carrot-like tuberous root and palmately lobed leaves. An annual shoot growing up to about 60 cm. tall is killed when the herb-layer is burnt annually. Only when the plant escapes burning for one year can the apical bud grow above the height of damage; then such plants can grow on into trees.

var. **bracteata** *Tennant* in K.B. 14: 223 (1960). Type: Tanganyika, Lindi District, Namanga [near Ruponda], *Anderson* 284 (EA, holo.!)

Rhachis of spike possessing overlapping, enlarged sterile bracts at base; flowers pentamerous.

Tanganyika. Lindi District: Namanga, 1 Jan. 1949 (fr.), *Anderson* 284!
Distr. **T**8; not known elsewhere
Hab. Deciduous woodland; 490 m.

Note. This variety, based on a single gathering, shows a mingling of characters of *C. zimmermannii* Harms and *C. kirkii* Seem. and is treated as a variety of the latter species because it possesses sessile fruits and a very close reticulum on the lower surface of the leaves; with the former species it has in common the possession of numerous enlarged sterile overlapping bracts at the base of the spike-rhachis. Further material is required.

var. **quadripetala** *Tennant* in K.B. 14: 223 (1960). Type: Tanganyika, Songea District, *Milne-Redhead & Taylor* 8274 (K, holo.!)

Flowers nearly always with 4 petals and 4 stamens; rhachis of spike closely similar to var. *kirkii*.

Tanganyika. Songea District: about 32 km. E. of Songea by R. Mkurira, 19 Jan. 1956 (fl. & young fr.), *Milne-Redhead & Taylor* 8274!
Distr. **T**8; not known elsewhere
Hab. Riverine forest; 930 m.

Note. Further material from this locality or elsewhere would be welcome.

5. **C. zimmermannii** *Harms* in E.J. 53: 361 (1915); T.T.C.L.: 59 (1949); K.T.S.: 54 (1961). Types: Tanganyika, Lushoto District, Amani, Kulemusi, *Zimmermann* in *Herb. Amani* 1042 (B, syn. †, EA, isosyn.!) & Amani, Cocosberg, *Zimmermann* in *Herb. Amani* 3042A (B, syn. †, EA, isosyn.!) & without precise locality, *Koerner* in *Herb. Amani* 2227 (B, syn. †, EA, isosyn.!) & Uzaramo District, without precise locality, *Stuhlmann* 6631 (B, syn. †)

A tree to 45 m. tall with greenish-grey fissured bark. Leaves digitately compound; petiole up to 53·5 cm. long and 5 mm. diameter, but generally considerably smaller, glabrous apart from some crisped hairs in stipular region and at junction with leaflets; leaflets 5–7(–9), sessile, chartaceous to coriaceous, lanceolate to oblanceolate, narrowly ovate and narrowly obovate, up to 25 cm. long by 8 cm. wide, but generally considerably less, caudate to acute (very rarely emarginate), with a narrowly cuneate to much attenuated base (occasionally base only a winged petiolule of up to 12 mm. long), with crenate to subentire margins, glabrous above, glabrous to slightly puberulous beneath. Flowering spikes up to about 12 together, up to 34 cm. long; floral bracts lanceolate to broadly ovate, up to 3·5 mm. long by up to 2 mm. wide, frequently much smaller, puberulous, sometimes glabrescent quite early on; base of spike-rhachis often clothed with larger, thicker overlapping sterile bracts. Fruits up to 6 mm. long, usually obconical to hemispherical, glabrous or puberulous. Fig. 3/5, p. 10.

Kenya. Kwale District: Shimba Hills, Mwele Mdogo Forest, 10 Feb. 1953 (fl.), *Drummond & Hemsley* 1198!; Tana River District: Mambosasa, Feb. 1929 (fl. & young fr.), *R. M. Graham* in *F.D.* 1806!
Tanganyika. Tanga District: Magunga Estate, 24 Jan. 1953 (fl. & fr.), *Faulkner* 1125!; Uzaramo District: Pugu Hills, 19 Mar. 1939 (fr.), *Vaughan* 2776!; Lindi District: Rutambo, 23 Mar. 1943 (fr.), *Gillman* 1350!
Distr. **K**7; **T**3, 6, 8; ? **Z**; NE. Mozambique
Hab. ? Lowland rain-forest, lowland dry evergreen forest, woodland; 0–400 m.

Variation. *Gillman* 1350 has more urceolate fruits than is usual.

NOTE. There is little doubt that the glabrous digitately compound leaf said to have been collected by Kirk in Zanzibar (see Hiern in F.T.A. 3: 32) but probably coming from the so-called Zanzibar Coast on the mainland, is referable to the present species. There is no evidence that this tree does in fact grow in Zanzibar.

6. **C. holstii** *Engl.* in Abh. Preuss. Akad. Wiss. 1894: 64 (1894); Harms in E. & P. Pf. 3 (8): 54 (1894) & in P.O.A. C: 298 (1895); Lebrun in B.J.B.B. 13: 17 (1934); T.S.K.: 115 (1936); F.P.N.A. 1: 691 (1948); T.T.C.L.: 59 (1949); I.T.U., ed. 2: 34 (1952); Tennant in K.B. 14: 223 (1960); K.T.S.: 52 (1961). Types: Tanganyika, Usambara Mts., " Mbindi " [probably near Mlalo], *Holst* (B, holo. †); Mlalo, *Semsei* 2829 (K, neo. !, EA, isoneo.)

A tree to 20 m. tall with a straight bole sometimes exceeding 1 m. in diameter and 10 m. or more tall in well grown specimens; bark rather fissured, shed in oblong papery scales. Leaves digitately compound; petiole up to 41·5 cm. long, glabrous or virtually so; lamina up to 36 cm. wide by about half as long; leaflets 3–7, chartaceous, ovate, up to 18·5 cm. long by 9 cm. wide, acuminate to caudate, with a cuneate to cordate, symmetric or asymmetric base, with serrate to crenate or entire margin, glabrous or puberulous, often with very fine silky-white scattered hairs and sometimes a few short crisped hairs; petiolules up to 7 cm. long; juvenile leaves (fig. 2/3, 4) simple, palmately lobed. Flowering spikes up to 30 together, usually less than 15, up to 25 cm. long, dense or somewhat lax; floral bracts scale-like, with or without an apiculum, or peg-like, nearly glabrous or densely pubescent. Fruits 4–6 mm. long, glabrous or puberulous, sometimes galled (see note on p. 1).

var. **holstii**; Tennant in K.B. 14: 224 (1960)

Rhachis of spike puberulous to pubescent; floral bracts scale-like, as broad at base as long, about 1 mm. long, with a rounded to acute apex, almost glabrous. Fruits urceolate with strongly convex lateral walls, 4·0–5·5 mm. long, not appreciably ribbed when dry; style-arms up to 0·5 mm. long in mature fruits, rarely widely divergent. Fig. 2/3–5.

UGANDA. Karamoja District: Kaabong, 13 Nov. 1939 (sterile), *A. S. Thomas* 3348 !; Kigezi District: between Kisoro and Lake Mutanda, Oct. 1940 (fl.), *Eggeling* 4131 !
KENYA. Northern Frontier Province: Moyale, 1 Nov. 1952 (sterile), *Gillett* 14111 !; Nakuru District: eastern Mau Forest Reserve, 9 Sept. 1949 (fr.), *Maas Geesteranus* 6207 !; Masai District: Cis-Mara, Feb. 1957 (fr.), *Ivens* 872 !
TANGANYIKA. Mwanza District: Bukumbi, Usagara, 27 Sept. 1952 (sterile), *Tanner* 1025 !; W. Kilimanjaro, small valley W. of Sanya sawmills, 13 Feb. 1951 (fr.), *Hughes* 54 !; Lushoto District: Mtowanguuwe, 28 Nov. 1958 (young fr.), *Semsei* 2817 !
DISTR. **U**1, 2; **K**1, 3, 4, 6; **T**1–3; Congo and Rwanda Republics, Ethiopia and Somali Republic (N.)
HAB. Upland dry evergreen forest, grouped-tree grassland, semi-evergreen bushland; 1110–2460 m.

SYN: *C. bequaertii* De Wild., Pl. Bequaert. 2: 89 (1923). Type: Congo Republic, Kivu Province, Angi, *Bequaert* 5791 (BR, holo.)
Prob. *Cussonia sp.* sensu Battiscombe, Cat. Trees Kenya Col.: 86, photo. (1926)
C. boranensis Cufod. in Miss. Biol. Borana, Racc. Bot., Angiosp.-Gymnosp.: 149, fig. 42 (1939). Type: Ethiopia, Galla-Sidamo, Araro [Arero], *Cufodontis* 284 (FI, holo.)

VARIATION. The gradation in leaf-form, from the simple palmately lobed juvenile leaves to the digitately compound leaves with petiolulate leaflets of mature branches, can be confusing. This variation in juvenile and mature foliage has not so far been observed in var. *tomentosa*, but the possibility of its occurrence should not be overlooked.

var. **tomentosa** *Tennant* in K.B. 14: 224 (1960). Type: Tanganyika, Kondoa District, Simbo Hills, *B. D. Burtt* 806 (K, holo. !, BM, EA, iso. !)

Rhachis of spike tomentose; floral bracts up to 2 mm. long, apiculate, densely tomentose to pubescent. Fruits urceolate, 5–6 mm. long, somewhat ribbed when dry, but lateral walls not strongly convex.

Fig. 2. *CUSSONIA SPICATA*—**1,** single leaflet, × ½; **2,** small leaf (complete), × ⅙. *C. HOLSTII* var. *HOLSTII*—**3, 4,** leaves from young parts of tree, × ⅙; **5,** leaf from mature part, × ⅙. 1, from *Drummond & Hemsley* 2469; 2, from *Bally* 7965; 3, 4, from *Gillett* 14111; 5, from *Gillett* 12899.

FIG. 3. *CUSSONIA LUKWANGULENSIS*—**1,** whole inflorescence terminating a leafy branch, × $\frac{1}{3}$; **2,** one of numerous primary branches of inflorescence, × 1. *C. SPICATA*—**3,** pedunculate spikes of flowers, × $\frac{1}{3}$. *C. JATROPHOIDES*—**4,** portion of one of several lax spikes of flowers forming the inflorescence, × $\frac{1}{2}$. *C. ZIMMERMANNII*—**5,** portion of one of several lax spikes of flowers forming the inflorescence, × $\frac{1}{2}$. 1, 2, from *E. M. Bruce* 769; 3, from *E. M. Bruce* 782; 4, from *B. D. Burtt* 5428; 5, from *Faulkner* 1125.

forma **tomentosa**; Tennant in K.B. 14: 224 (1960)

Leaflets with a serrate to crenate margin. Style-arms approximately 1 mm. long in mature fruits, delicate, often widely divergent.

TANGANYIKA. Singida District: Rift Wall, near Maw Hills, Oct. 1935 (young fr.), *B. D. Burtt* 5230!; Kondoa District: Simbo Hills, 14 Dec. 1927 (fr.), *B. D. Burtt* 806!; Dodoma District: Manyoni Kopje, 21 Nov. 1931 (fr.), *B. D. Burtt* 3418!
DISTR. **T5**; not known elsewhere
HAB. *Brachystegia microphylla* woodland, often in rocky places; 1300–1700 m.

forma **integrifoliola** *Tennant* in K.B. 14: 224 (1960). Type: Tanganyika, Shinyanga District, Usanda, *Koritschoner* 1850 (K, holo. !, EA, iso. !)

Leaflets with an almost entire margin. Style-arms up to 0·5 mm. long in mature fruits.

TANGANYIKA. Shinyanga District: Usanda, *Koritschoner* 1850!
DISTR. **T1**; not known elsewhere
HAB. Not known

NOTE. The material to hand of the two forms of var. *tomentosa* is most scanty, and forma *integrifoliola* is based on a single gathering. Until more material is available, it seems expedient to retain a single species for the whole assemblage, separating the uniform var. *holstii* from the less homogeneous var. *tomentosa*.

7. **C. lukwangulensis** *Tennant* in K.B. 14: 221 (1960). Type: Tanganyika, Uluguru Mts., Tarana, *E.M. Bruce* 769 (K, holo. !, BM, iso. !)

A much branched tree to 15 m. tall (*fide* Greenway & Eggeling), or an epiphyte on trees (*fide* E.M. Bruce). Leaves digitately compound with up to 4 shortly petiolulate leaflets, or sometimes reduced to a single similar leaflet; petiole up to 18 cm. long and 1·5 mm. diameter, ribbed, glabrous and uniform in cross-section for greater part of length, expanding considerably at base; leaflets narrowly oblanceolate to narrowly oblong-oblanceolate, very narrowly elliptic and very narrowly oblong-elliptic, up to 14·5 cm. long by 3·3 cm. wide, long-acuminate, attenuate at base, with the margin slightly inrolled, entire to repand and slightly crisped, glabrous or nearly so, often with circular pustules of up to 0·3 mm. diameter covering both surfaces; petiolules up to 2·2 cm. long. Inflorescence a group of up to about 20 racemes of umbellules, each (raceme) up to 20 cm. long and floriferous only on the terminal third, in which region the peduncles of umbellules are arranged spirally; peduncles 1·5–2·5 cm. long; umbellules 10–15-flowered; pedicels up to 7 mm. long. Fig. 3/1, 2.

TANGANYIKA. Uluguru Mts., Lukwangule Plateau, 23 Jan. 1933 (fl.), *Schlieben* 3566! & Morogoro–Lupanga Peak track, 16 Aug. 1951 (young fr.), *Greenway & Eggeling* 8610!
DISTR. **T6**; known only from the Uluguru Mts.
HAB. Upland rain-forest on steep slopes, locally dominant on rocky ridges; 1380–2400 m.

SYN. ? *C. buchananii* Harms in E.J. 26: 251 (1899) pro parte, quoad spec. Tanganyika, Uluguru Mts., Lukwangule Plateau, *Stuhlmann* 9112 (B, syn. †)

NOTE. This species is easily distinguished from the allied *C. umbellifera* Sond. var. *buchananii* (Harms) Tennant, known from Malawi, by the longer acuminate apices, and more attenuate bases of its leaflets, which in addition are generally considerably narrower. It is hoped that further well annotated material of *C. lukwangulensis* will reveal whether the epiphytic habit observed by Miss E. M. Bruce is widespread.

2. POLYSCIAS

J.R. & G. Forst., Char. Gen.: 63, t. 32 (1776); emend. Harms in E. & P. Pf. 3(8): 43 (1894)

[*Panax* sensu Hiern in F.T.A. 2: 27 (1877), *non* L.]
[*Sciadopanax* sensu R. Viguier in Bull. Soc. Bot. Fr. 52: 304 (1905), *non* Seem.]

Trees. Leaves petiolate, more than once pinnately compound or pari- or imparipinnate, glabrous or pubescent with simple or stellate hairs. Leaflets simple with margin generally either entire or nearly so, or sometimes pinnately compound or pinnatisect. Stipules present or absent. Inflorescences paniculate, glabrous to densely tomentose, sometimes branching in a very orderly racemose manner (species 1–3); ultimate branches (pedicels) arranged in racemules or umbellules; pedicels (except in *P. stuhlmannii* var. *inarticulata*) articulate just below the flower. Flowers with calyx-margin repando-5-dentate to subentire. Petals 5. Stamens 5. Disk nearly flat, slightly convex or conically raised. Ovary bilocular or (in *P. stuhlmannii*) 5-locular; styles 2, connivent at base, or (in *P. stuhlmannii*) 5, free or nearly so. Fruit laterally compressed or practically terete. Seeds laterally compressed, ellipsoidal, fluted or smooth; endosperm non-ruminate.

A widespread genus in the tropics of the Old World, containing just over 100 species, or possibly less after monographic revision.

NOTE. The length of fruits in the following descriptions excludes the stylopodium.

1. Flowers in racemules (branching of inflorescence entirely racemose), see fig. 4/2; fruits with 2 persistent styles; leaflets densely stellate-tomentose beneath . .	1. *P. fulva*
Flowers in umbellules, see fig. 4/8; fruits with 2 or 5 persistent styles; leaflets tomentose or glabrous	2
2. Ovary 5-locular; fruits with 5 styles; leaves glabrous on both surfaces . . .	4. *P. stuhlmannii*
Ovary 2-locular; fruits with 2 styles	3
3. Inflorescence-branches and young fruits densely stellate-tomentose, mature fruits and inflorescence-branches more sparsely tomentose	2. *P. kikuyuensis*
Inflorescence-branches and both young and mature fruits glabrous	3. *P. albersiana*

1. **P. fulva** (*Hiern*) *Harms* in E. & P. Pf. 3 (8): 45 (1894); V.E. 3 (2): 780 (1921); F.P.N.A. 1: 692 (1948); T.T.C.L.: 60 (1949); I.T.U., ed. 2: 34, photo. 3 (1952); F.P.S. 2: 356 (1952); F.W.T.A., ed. 2, 1: 750 (1958); F.F.N.R.: 313 (1962). Type: Fernando Po, *Mann* 301 (K, holo. !)

A tree to 30 m. tall, often with a large grey extremely straight unbranched cylindrical bole of up to 15 m. tall and 1 m. diameter, generally dividing into a small number of main branches which themselves each branch in a similar manner. Leaves up to 80 cm. long, generally imparipinnate, less often paripinnate; leaflets (3–)6–7(–12) pairs, chartaceous to coriaceous, lanceolate to ovate (occasionally very broadly ovate), often markedly straight-edged, up to 14(–17) cm. long and up to 5·5(–7·5) cm. wide, acute to acuminate, generally apiculate, rarely retuse, rounded or obtuse, with a subcordate, truncate or rounded, rarely very broadly cuneate, occasionally slightly oblique base, and entire occasionally slightly undulate, generally very narrowly inrolled margins, densely stellate-tomentose when young, irregularly glabrescent to some degree (especially above) later; petiolules of paired leaflets (0–)5(–14) mm. long. Inflorescence-branches of three orders (fig. 4/2), racemosely borne, pubescent to tomentose, often irregularly glabrescent, sometimes galled (see p. 1); primaries up to 70 cm. long, 3·0–6·5 mm. diameter; secondaries up to 3–7(–12) cm. long, 0·7–1·7(–2·0) mm. diameter; tertiaries (pedicels) up to 4 mm. (not exceeding 2·5 mm. in Flora area) long, 0·3–0·8(–1·0) mm. diameter. Flowers greenish-yellow to cream; styles 2, persistent in fruit. Fruits ovoid to obovoid, occasionally ellipsoid, shortly

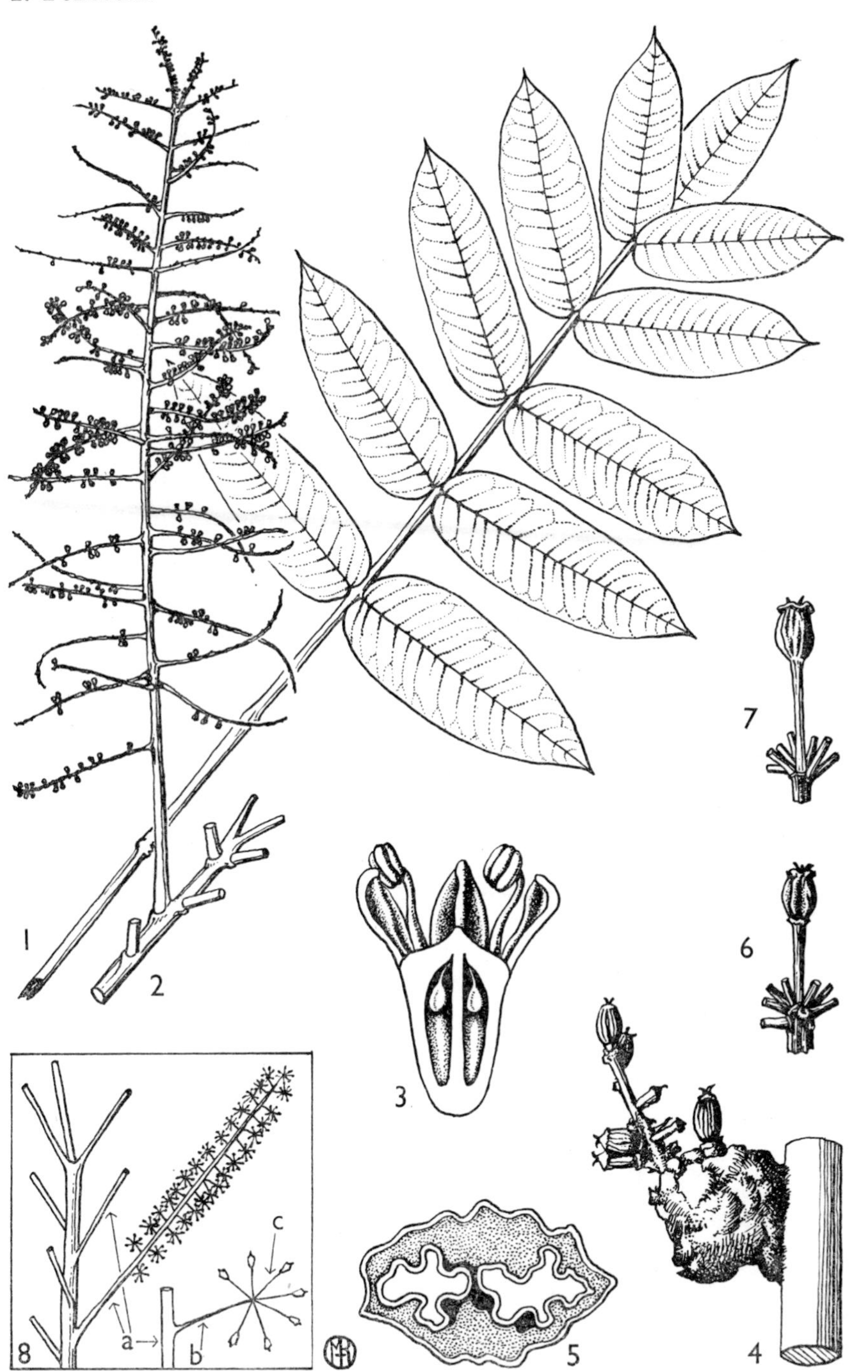

FIG. 4. *POLYSCIAS FULVA*—**1,** leaf (some of the lower pinnae lacking), × $\frac{1}{3}$; **2,** inflorescence-axis and one primary branch of inflorescence, × $\frac{1}{3}$; **3,** vertical section of flower, × 12; **4,** portion of a primary branch of inflorescence showing galled secondary branch, × 2; **5,** transverse section of fruit, × 13. *P. STUHLMANNII* var. *STUHLMANNII*—**6,** mature fruit with articulated pedicel, × 1. *P. STUHLMANNII* var. *INARTICULATA*—**7,** mature fruit with inarticulated pedicel, × 1. *P. SPP.*—**8,** theoretical diagrams of inflorescence (much reduced) relating to both *P. albersiana* and *P. kikuyuensis*, showing main axis bearing primary branches of inflorescence (a), which in turn bear pedunculate umbellules of flowers; b = secondary branches (peduncules of umbellules); c = tertiary branches (pedicels). 1, from *R. M. Davies* 881; 2, 3, from *Drummond & Hemsley* 1752; 4, 5, from *Sangster* 524; 6, from *Drummond & Hemsley* 4350; 7, from *Eggeling* 6668.

cylindroidal or subspheroidal, 3·5–6·0 mm. long by 3·0–4·5 mm. maximum width, generally ribbed and markedly flattened, glabrous or very sparsely stellate-hairy. Fig. 4/1–5, p. 13.

UGANDA. E. Ruwenzori, May 1939 (young fr.), *Sangster* 524!; Elgon, Feb. 1940 (fr.), *St. Clair-Thompson* in *Eggeling* 3953!; Masaka District: Sese Is., Bugala I., Nov. 1931 (fl.), *Eggeling* 99!

KENYA. N. Kavirondo District: Kakamega, June 1934 (fl.), *Dale* in *F.D.* 3254!

TANGANYIKA. Bukoba District: Nyakato, 14 Apr. 1935 (fr.), *Gillman* 222!; W. Usambara Mts., Bumbuli–Mazumbai road, 9 May 1953 (fl.), *Drummond & Hemsley* 2495!; Rungwe District: Mt. Rungwe and Poroto Mts., 11 Mar. 1932 (fl.), *St. Clair-Thompson* 982!

DISTR. **U**2–4; **K**5, ? 7*; **T**1–4, 6, 7; widely spread from Guinée Republic to Ethiopia, and southwards through the Congo Republic to Malawi, Zambia, Rhodesia and Angola

HAB. Upland and lowland rain-forest, riverine forest, also upland grassland; 1180–2160 m.

SYN. *Panax fulvum* Hiern in F.T.A. 3: 28 (1877)
P. ferrugineum Hiern in F.T.A. 3: 28 (1877). Type: Ethiopia, Shoa, Ankober, *Roth* (K, holo.!)
Polyscias ferruginea (Hiern) Harms in E. & P. Pf. 3 (8): 45 (1894); K.T.S.: 54 (1961)
P. preussii Harms in E.J. 26: 245 (1899). Type: Cameroun Republic, Bouea [Buea], *Preuss* 887 (B, holo. †, BM, iso.!)
P. elliotii Harms in E.J. 26: 246 (1899). Type: Uganda, Ruwenzori, without precise location, *Scott Elliot* 7766 (whereabouts uncertain)
P. polybotrya Harms in N.B.G.B. 3: 20 (1900). Type: Tanganyika, E. Usambara Mts., Ngwelo–Derema, *Scheffler* 53 (B, holo. †, BM, EA, iso.!)
P. malosana Harms in N.B.G.B. 3: 20 (1900). Type: Malawi, Mt. Malosa, *Whyte* (B, holo. †, ? K, iso.!)
Sciadopanax ferruginea (Hiern) R. Viguier in Bull. Soc. Bot. Fr. 52: 304 (1905)
S. fulva (Hiern) R. Viguier in Bull. Soc. Bot. Fr. 52: 305 (1905)
S. preussii (Harms) R. Viguier in Bull. Soc. Bot. Fr. 52: 305 (1905)
S. elliotii (Harms) R. Viguier in Bull. Soc. Bot. Fr. 52: 305 (1905)
S. polybotrya (Harms) R. Viguier in Bull. Soc. Bot. Fr. 52: 305 (1905)
S. malosana (Harms) R. Viguier in Bull. Soc. Bot. Fr. 52: 305 (1905)
Panax nigericum A. Chev. in Mém. Soc. Bot. Fr. 8: 178 (1912). Type: Guinée Republic, Timbikounda, Farana, *Chevalier* 20590 (P, holo., K, iso.!)

VARIATION. The cordate or non-cordate leaflet-base key-character employed by Hiern to separate his *Panax ferrugineum* and *P. fulvum* and later employed by Harms to distinguish *Polyscias malosana* from *P. elliotii*, and again one of the characters used by A. Chevalier to distinguish *Panax nigericum* from *P. fulvum*, cannot be upheld, and the same conclusion is reached here as that drawn by Keay in F.W.T.A., ed. 2. Pedicel-length has in the past been used to separate some of the species in the present synonymy, but this is shown to be of no specific importance; thus the distinction drawn by Harms between *Polyscias elliotii*, *P. preussii* and *P. polybotrya* and later by Chevalier between his *Panax nigericum*, *P. fulvum* and *P. ferrugineum* on the basis of discontinuities in this character cannot be upheld.

2. **P. kikuyuensis** *Summerh.* in K.B. 1926: 242 (1926); Battiscombe, Cat. Trees Kenya Col.: 86, photo. (1926); T.S.K.: 114 (1936); T.T.C.L.: 60 (1949); K.T.S.: 55 (1961). Type: Kenya, "[Kiambu District,] Kikuyu Escarpment and [Nakuru District,] Elburgon Forests," *Cooper* in *Battiscombe* 873 (K, holo.!, EA, iso.!)

A tree to 25 m. tall, often with an unbranched bole up to 12 m. tall and 1·2 m. diameter. Leaves up to 55 cm. long, imparipinnate, less frequently paripinnate; leaflets (3–)4–5(–6) pairs, coriaceous, rarely chartaceous, lanceolate to narrowly ovate, very occasionally rotund, often straight-edged or oblique, up to 14·5 cm. long by 6·5 cm. wide (or larger in saplings), acute

* A sterile specimen of sapling leaves, *Drummond & Hemsley* 4378! from Kenya, Teita Hills, Bura Bluff, Chawia Forest, with leaflets up to 23·5 cm. long by 15 cm. wide, should perhaps be included under the present species, but since the sapling leaves of *P. fulva* and *P. kikuyuensis* are indistinguishable, the name of the specimen must remain in doubt. This specimen represents a new record for **K**7 for either species.

to acuminate, rarely emarginate, with a rounded to cordate (often subcordate) base, with the margins entire, often very narrowly inrolled, densely stellate-tomentose when young, later glabrescent to some extent, especially above where occasionally glabrous; petiolules of paired leaflets (0–)2–5(–14) mm. long. Inflorescence-branches puberulous to tomentose; primaries up to 40 cm. long by 2·5–4·0 mm. diameter; secondaries up to 2·7 cm. long by 0·8–1·2 mm. diameter, both orders racemosely borne; tertiaries (pedicels) up to 9 mm. (commonly 2–5 mm.) long by 0·5–0·8 mm. diameter in umbellules. Fruits flattened-ovoid, -ellipsoid, -cylindrical or ± spheroidal, 4·0–7·5 mm. long, ribbed, apart from the stylopodia and persistent styles densely (often interruptedly) covered with stellate hairs. Fig. 4/8, p. 13.

KENYA. SE. Aberdare Mts., 31 Dec. 1926 (fr.), *Hoult* 93!; Nairobi or Masai District: Bahati Forest, Jan. 1932 (fr.), *Afford* in *F.D.* 517, 9075!; Meru District: Lari [Lare], *C. F. Elliott* in *C.M.* 14600!
DISTR. **K**3, 4, ? 6; not known elsewhere
HAB. Upland rain-forest; 1750–2620 m.

SYN. [*P. fulva* sensu Chiov., Racc. Bot. Miss. Consol. Kenya: 50 (1935), pro parte, *non* (Hiern) Harms]

NOTE. The leaflets of the type-gathering, *Cooper* in *Battiscombe* 873, are up to about 18 cm. long by 10·5 cm. wide and well beyond the range of other material of the species seen; they are almost certainly from sapling shoots (see footnote under *P. fulva*, p. 14).
Segmental pupae are sometimes found in the cocci of fruit, in which cases the endosperm is largely disintegrated and much reduced in amount.

3. **P. albersiana** *Harms* in E.J. 33: 182 (1902); T.T.C.L.: 60 (1949). Type: Tanganyika, W. Usambara Mts., Kwai, *Albers* 317 (B, holo. †)

A tree to 20 m. tall with grey bark and a flattish crown. Leaves up to 60 cm. long (sometimes more than double this length on sapling branches), imparipinnate; leaflets 7–11 pairs, chartaceous to coriaceous, lanceolate to ovate or broadly elliptic, frequently ± oblique, up to 13 cm. long by 5·5 cm. wide (sometimes nearly twice as large on saplings), acuminate, cordate to rounded at base, entire to shallowly undulate, stellate-pubescent beneath and sparsely so above when young, glabrescent; petiolules of paired leaflets on mature branches up to 0·8 cm. long, glabrous, but up to 3·5 cm. long and pubescent on sapling branches. Inflorescence-branches glabrous or almost so; primaries up to 30 cm. long by 3 mm. diameter; secondaries 1·0–4·5 cm. long by 0·5–1·0 mm. diameter, both orders racemosely borne; tertiaries (pedicels) 3–15 mm. long by 0·3–0·6 mm. diameter in umbellules. Flowers greenish-yellow. Styles 2, persistent. Fruits flattened-ovoid to -ellipsoid, up to 6·5 mm. long, ribbed, glabrous. Fig. 4/8, p. 13.

TANGANYIKA. Kilimanjaro, Aug. 1912 (fl.), *Naepfel* 2968 (*fide* Harms); W. Usambara Mts., Lushoto–Malindi road, saddle near Magamba Peak, 29 May 1953 (fr.), *Drummond & Hemsley* 2807! & 2807a! & Magamba–Shume road, 2 Apr. 1923 (fl.), *Rawcliffe* 58!; Ufipa District: Sumbawanga, Kawa R., 30 Dec. 1956 (fl.), *Richards* 7413B!
DISTR. **T**2–4; not known elsewhere
HAB. Upland rain-forest; 1250–2000 m.

SYN. *Sciadopanax albersiana* (Harms) R. Viguier in Bull. Soc. Bot. Fr. 52: 305 (1905)
Polyscias albersii Engl. in V.E. 3 (2): 779 (1921), error for *P. albersiana* Harms

NOTE. No authentic material of this species has been seen; the above description reconciles the original description by Harms with the characters of the four specimens studied for this work, some of which come from near the type-locality.

4. **P. stuhlmannii** *Harms* in E.J. 26: 244 (1899); V.E. 3 (2): 779 (1921); T.S.K.: 114 (1936); T.T.C.L.: 60 (1949); K.T.S.: 55 (1961). Type: Tanganyika, Uluguru Mts., Lukwangule plateau, *Stuhlmann* 9122 (B, holo. †)

A tree to 20 m. tall with grey or whitish bark and often with a large spreading crown, or a large spreading shrub (*fide* Gardner). Leaves up to 40–60 cm. long, often less, almost always imparipinnate, with no clear-cut differentiation between juvenile sapling leaves and leaves from mature branches, glabrous; leaflets 4–5 pairs, coriaceous, narrowly ovate to obovate, up to 12 cm. long by 5·5 cm. wide, commonly only about half these dimensions, obtuse to minutely retuse, occasionally acute or with a broad rounded apiculum, broadly cuneate to attenuate at base, entire to repand and narrowly inrolled at the margin; petiolules of lateral leaflets 4(–8) mm. long. Inflorescence-branches glabrous (occasionally somewhat tuberculate); pedicels umbellulately borne (occasionally flowers borne singly); other orders of branching may be umbellate or racemose. Petals dark reddish-brown to purple-black. Styles 5, persistent. Fruits urceolate, 7–9 mm. long, terete, deeply 5-sulcate when mature, glabrous.

var. **stuhlmannii**

Pedicels articulated just beneath the flower. Fig. 4/6, p. 13.

KENYA. Teita District: 8 km. NNE. of Ngerenyi, Ngangao, 15 Sept. 1953 (fr.), *Drummond & Hemsley* 4350! & Ngangao Hill, July 1937 (young fr.), *Dale* in *F.D.* 1130! & Sept. 1932 (fl. & fr.), *Gardner* in *F.D.* 2922!

TANGANYIKA. W. Usambara Mts., Shagai Forest near Sunga, 17 May 1953 (fl. & young fr.), *Drummond & Hemsley* 2606! & Lushoto–Malindi road, near Magamba Peak, 29 May 1953 (fr.), *Drummond & Hemsley* 2810!; Uluguru Mts., Lupanga, 7 Apr. 1935 (fr.), *E. M. Bruce* 998!

DISTR. **K**7; **T**3, 6; not known elsewhere

HAB. Upland rain-forest; 1700–2300 m.

SYN. *Gastonia stuhlmannii* (Harms) Harms in E.J. 53: 360 (1915), pro parte

var. **inarticulata** *Tennant* in K.B. 14: 400 (1960). Type: Tanganyika, W. Usambara Mts., Shume–Magamba Forest Reserve, Gologolo, *Eggeling* 6668 (K, holo. !, EA, iso. !)

Pedicels always or nearly always inarticulate. Fig. 4/7, p. 13.

TANGANYIKA. W. Usambara Mts., Shume–Magamba Forest Reserve, 24 Nov. 1952 (fr.), *Parry* 189! & Magamba Forest Reserve, Nov. 1957 (fr.), *Semsei* 2711!

DISTR. **T**3 (known only from the W. Usambara Mts.)

HAB. Upland rain-forest; 1700–2060 m.

SYN. *Gastonia stuhlmannii* (Harms) Harms in E.J. 53: 360 (1915), pro parte quoad specim. W. Usambara Mts., Magamba, *Deininger* 2883 (B, †)

NOTE. Although pedicels are almost all lacking articulation, traces of the articulation are still evident in a few of the younger fruits and buds of *Eggeling* 6668. The lack or possession of this character (which alone distinguishes the two varieties of the present species) is thus possibly not of such importance in the classification of genera within this family as it was formerly thought to be. So far, var. *inarticulata* is known only from the above cited specimens and more material, accompanied by careful field notes, is required.

3. SCHEFFLERA

J.R. & G. Forst., Char. Gen.: 45, t. 23 (1776); emend. Harms in E. & P. Pf. 3 (8): 35 (1894), *nom. conserv.*

Heptapleurum Gaertn., Fruct. & Semin. Plant. 2: 472, t. 178, fig. 3 (1791); Hiern in F.T.A. 3: 29 (1877)

Trees or lianes, often epiphytic or possibly epiphytic at first and independent later. Leaves petiolate, digitately compound; leaflets subentire to crenate. Stipules intrapetiolar, often connate with petiole for some distance. Inflorescences paniculate, with flowers ultimately in umbellules, fascicles or capitula; pedicels (where present) not articulated beneath the flower. Calyx-margin repando-dentate to subentire. Petals 5–10, sometimes calyptrate (being shed as a cup) or opening on flower. Stamens equal in number to the petals. Disk flat, depressed or conically raised. Ovary 5–8-locular;

styles equal in number to the locules, connivent at base. Fruit ± subglobose or ovoid. Seeds laterally compressed, ellipsoidal, smooth; endosperm non-ruminate.

A widely distributed Old and New World genus, with the majority of the several hundred species Indo-Malesian.

NOTE. The length of fruits in the following descriptions excludes the stylopodium.

1. Flowers pedicellate in umbellules (sect. *Schefflera*) . . . 2
   Flowers sessile in pedunculate capitula (sect. *Cephaloschefflera* Harms), see fig. 5/8 4
2. Inflorescence (fig. 5/9) with entirely or predominantly umbellate branching; peduncles of umbellules all or nearly all umbellulately borne; leaflets with many close lateral veins . . 3. *S. polysciadia*
   Inflorescence with racemose and umbellate branching; peduncles of umbellules racemosely borne; lateral veins of leaflets fewer and further apart 3
3. Lateral veins of leaflets dividing ± midway between midrib and margin into two or more major branches . . . . 1. *S. barteri*
   Lateral veins of leaflets dividing only near the margin or not at all . . . . . 2. *S. abyssinica*
4. Leaflets acute to rounded at apex; petiolules up to 2·1(–2·8) cm. long; peduncles of capitula (0·5–)1(–1·7) cm. long . . . . . 4. *S. volkensii*
   Leaflets long-acuminate; petiolules 2·5–6 cm. long; peduncles of capitula up to 0·5(–0·9) cm. long 5. *S. stolzii*

1. **S. barteri** (*Seem.*) *Harms* in E. & P. Pf. 3 (8): 38 (1894); V.E. 3 (2): 778 (1921); Lebrun in B.J.B.B. 13: 21 (1934); Exell., Cat. Vasc. Pl. S. Tomé: 194 (1944); F.W.T.A., ed. 2, 1: 751 (1958); Tennant in K.B. 15: 331 (1961). Type: Sierra Leone, Sugar Loaf Mt., *Barter* 2027 (K, holo.!)

A woody liane to 30 m. or more tall, an erect tree or a large or low scrambling shrub. Petioles up to 57 cm. long by 9 mm. diameter at base, ribbed, glabrous or discontinuously puberulous; leaflets 4–11, chartaceous to coriaceous, narrowly to broadly elliptic, oblanceolate to narrowly obovate, lanceolate to narrowly ovate, sometimes somewhat oblong, up to 25(–29) cm. long and 10(–12) cm. wide, apiculate or mucronate to long-acuminate, occasionally with a broad rounded apiculum, with a rounded to obtuse or narrowly cuneate base, subentire to shallowly crenate or serrulate, often obsoletely serrulate towards tip and nearly entire below, with 5–14 pairs of lateral veins*; petiolules 0·5–8(–10) cm. long by 0·5–2·5 mm. diameter. Stipules up to (and probably exceeding) 2·7 cm. long, generally partially adnate to petiole. Inflorescence (fig. 5/4) an umbel of 3–10(–20) racemes of umbellules; primary branches 9–45 cm. long, sterile at base, arising from a group of tomentose glabrescent bracts up to 4 cm. long, bearing sessile or pedunculate umbellules of flowers; peduncles of umbellules (secondary branches) where developed 0–3·2 cm. long, racemosely borne along the primaries, typically in the axils of tomentose glabrescent ± deltoid or

* The numbers of lateral veins takes into account only those which arise from the midrib and branch approximately midway to the margin giving rise to two (or sometimes more) well-defined branches which can be traced nearly to the margin. There are often other smaller veins arising from the midrib between those mentioned above; these however branch and anastomose, losing their identity well inside the margin.

FIG. 5. *SCHEFFLERA BARTERI* var. *BARTERI*—**1**, leaf, × ⅓; **2**, mature flower-bud, with corolla removed, × 12; **3**, calyptrate petals of same, × 12; **4**, raceme of fruiting umbellules—one of several primary branches of an infructescence, × ⅓; **5**, fruit, × 2½; **6**, vertical section of same, × 5; **7**, transverse section of same, × 10. *S. SPP.*—**8**, theoretical diagram of inflorescence (much reduced) relating to both *S. volkensii* and *S. stolzii*, showing the main axis bearing several primary branches (a), which in turn bear the racemosely arranged secondary branches (viz. peduncles of the capitula) (b); **9**, theoretical diagram of the inflorescence (much reduced) of *S. polysciada*, showing the main axis bearing several umbellately arranged primary branches (a), which in turn bear the racemosely and umbellately arranged secondary branches (b), each with tertiary branches (peduncles of umbellules) (c) and quaternary branches (pedicels) (d). 1, from *Drummond & Hemsley* 1618; 2, 3, from *Wallace* 456; 4–7, from *E. M. Bruce* 240.

lanceolate to ovate (occasionally obsolete) bracts* up to 7(–12) mm. long; pedicels up to about 24 per peduncle, often less, (2–)4–12 mm. long, subtended by minute or obsolete bracts. Fruits urceolate to hemiellipsoidal, up to 4·5 mm. long by 3·5(–4·5) mm. diameter, sulcate, glabrous to puberulous; stylopodium 0·5–1·0 mm. long; styles 5–8, free for up to 0·7 mm. terminally.

var. **barteri**; Tennant in K.B. 15: 331 (1961)

Secondary branches of inflorescence (peduncles of umbellules) (0·4–)0·7–3·2 cm. long. Fig. 5/1–7.

UGANDA. Ruwenzori, Namwamba valley, 3 Jan. 1935 (fl. & young fr.), *G. Taylor* 2809!
TANGANYIKA. Kilimanjaro above Marangu, 29 Dec. 1912 (fl. & fr.), *Grote* in *Herb. Amani* 4029!; W. Usambara Mts., Makuyuni, *Koritschoner* 1559!; Uluguru Mts., 5 km. S. of Bunduki, Salaza Forest, 15 Mar. 1953 (fr.), *Drummond & Hemsley* 1618!
DISTR. **U2**; **T2**, 3, 6, 8; Congo and Cameroun Republics and West Africa to Guinée Republic; also in Rhodesia and possibly Angola
HAB. Lowland or upland rain-forest; 900–2000 m.

SYN. *Astropanax barteri* Seem. in J.B. 3: 177 (1865)
Sciadophyllum barteri (Seem.) Seem. in J.B. 3: 267 (1865) & Rev. Hed.: 51 (1868)
Schefflera goetzenii Harms in von Goetzen, Durch. Afr.: 376 & 380 (1895) & in P.O.A. A: 134 (1895) & in von Goetzen, Durch. Afr. [reprint]: 7 (1896) & in E.J. 26: 242 (1899); V.E. 3 (2): 778 (1921); Lebrun in B.J.B.B. 13: 22 (1934); F.P.N.A. 1: 688 (1948); I.T.U., ed. 2: 35 (1952). Type: Congo Republic, between Lakes Edward and Kivu, Virunga Mts., *von Goetzen* 46 (B, holo. †). A probable synonym
S. stuhlmannii Harms in E.J. 26: 243 (1899); V.E. 3 (2): 778 (1921); T.T.C.L.: 61 (1949). Types: Tanganyika, Uluguru Mts., *Stuhlmann* 8849 & upper Mgata [? Mgeta] valley, *Stuhlmann* 9277 (both B, syn. †)
S. adolfi-friderici Harms in Z.A.E.: 590 (1913); V.E. 3 (2): 778 (1921); Lebrun in B.J.B.B. 13: 21 (1934); F.P.N.A. 1: 688 (1948). Type: Rwanda Republic, volcanic region of Bugoie, *Mildbraed* 1478 (B, holo. †). A probable synonym
S. mildbraedii Harms in Z.A.E.: 591 (1913); V.E. 3 (2): 779 (1921); Lebrun in B.J.B.B. 13: 21 (1934); F.P.N.A. 1: 688 (1948). Type: Rwanda Republic, Rugege Forest, Rukarara [stream], *Mildbraed* 1010 (B, holo. †). A probable synonym
S. sycifiifolia Lebrun in B.J.B.B. 13: 19 (1934); F.P.N.A. 1: 687 (1948). Type: Congo Republic, Kivu Province, Kasindi–Lubango, *Lebrun* 4755 (BR, holo., K, iso.!, BM, iso. fragment!)

NOTE. A wide survey of tropical African species has allowed the union of *S. urostachya* and *S. stuhlmannii* with the well-known West African species *S. barteri*. Distinction at varietal level based upon umbellule peduncle-length is still feasible, however, in the case of the two East African taxa. This character is absolute in many instances, only rarely with slight overlap between the varieties. The size of bracts subtending the peduncles is apparently of no taxonomic significance.

Distribution of both varieties in the present Flora is distinct as far as records indicate, except for **U2** where both have been collected.

var. **urostachya** (*Harms*) *Tennant* in K.B. 15: 333 (1961). Type: Congo Republic, Kivu Province, Beni near Muera, *Mildbraed* 2185 (B, holo. †)

Secondary branches of inflorescence (peduncles of umbellules) 0–0·6(–0·7) cm. long.

UGANDA. Kigezi District: Ishasha Gorge, Apr. 1946 (fl. & young fr.), *Purseglove* 2031!; Masaka District: Malabigambo Forest, 6 km. SSW. of Katera, 3 Oct. 1953 (fr.), *Drummond & Hemsley* 4609!; Mengo District: 16 km. on Kampala–Bombo road at extreme S. end of Semunya Forest, Jan. 1938 (fl. & young fr.), *Chandler* 2121!
TANGANYIKA. Bukoba District: Rubare Forest Reserve, Kaigi, May. 1935 (fl.), *Gillman* 271!
DISTR. **U2**, 4; **T1**; Rio Muni, S. Tomé, Gabon, Cameroun and Congo Republics
HAB. Fresh water swamp-forest, ? lowland rain-forest; 1140–1230 m.

* When the umbellules are to any degree pedunculate, the relation of the peduncle and its subtending bract is obvious; however, when the umbellules are sessile, the bract is not normally suppressed, but retains its theoretical position on the primary branch of the inflorescence, and consequently arises in very close proximity to the pedicels and their subtending bracts. On account of its considerably larger size and more foliaceous nature it is easily distinguished from the much smaller pedicel-associated bracts.

SYN. *S. urostachya* Harms in Z.A.E.: 591, t. 78 (1913); V.E. 3 (2): 779 (1921); Lebrun in B.J.B.B. 13: 22 (1934); F.P.N.A. 1: 687 (1948); T.T.C.L.: 61 (1949); I.T.U., ed. 2: 36 (1952)
S. tessmannii Harms in E.J. 53: 360 (1915); V.E. 3 (2): 779 (1921). Type: Rio Muni, Nkolentangan, *Tessmann* 344 (B, holo. †, K, iso.!)
S. sp. sensu Check-lists For. Trees & Shrubs Brit. Emp., No. 1, Uganda Prot.: 24 (1935) quoad spec. *Maitland* 414!, *Brown* 147!, *A. S. Thomas* 926!, *Dawe* 216!, *F.D.* 410!

2. **S. abyssinica** (*A. Rich.*) *Harms* in E. & P. Pf. 3 (8): 38 (1894); V.E. 3 (2): 778 (1921); Lebrun in B.J.B.B. 13: 20 (1934) & Ess. For. Rég. Mont. Congo Or.: 183 (1935); Check-lists For. Trees & Shrubs Brit. Emp., No. 1, Uganda Prot.: 24 (1935); T.S.K.: 116 (1936); F.P.N.A. 1: 688 (1948); T.T.C.L.: 61 (1949); I.T.U., ed. 2: 35 (1952); F.P.S. 2: 357 (1952); F.W.T.A., ed. 2, 1: 751 (1958); T.S.K.: 55 (1961); F.F.N.R.: 313 (1962). Type: Ethiopia, Adua, Mt. Scholoda, *Schimper* 283 (BM, K, isosyn.!)

An often spreading tree to 28 m. tall or an epiphyte; bark rough and fissured or smooth, grey-brown to grey-black and ± corky. Petioles up to 38(–42) cm. long by 5(–6·2) mm. diameter, glabrous or sparsely hairy at tip, rarely puberulous at base; leaflets 5–7, chartaceous to coriaceous, elliptic to broadly elliptic, narrowly ovate to ovate, sometimes somewhat oblong, occasionally narrowly obovate to obovate or oblanceolate, up to 27(–40) cm. long by 15(–20) cm. wide, gradually tapering to a ± truncate short- to long-acuminate or almost caudate apex, with a cordate or rounded, obtuse to cuneate base, with crenulate (rarely subentire) margins, glabrous or puberulous; petiolules up to 12·5 cm. long by 1–3 mm. diameter. Stipules up to 1·7 cm. long. Inflorescence an umbel or a short raceme of up to 12 racemes of pedunculate umbellules; primary branches up to 41 cm. long, often not floriferous at base, arising in axils of tomentose to puberulous bracts of up to 1·5 cm. long (this relationship obscured when primary branches umbellate); peduncles of umbellules (secondary branches) (0·6–)1·3–2·1(–4·5) cm. long, arising in the axils of glabrous to puberulous ovate to ligulate bracts of from 1·5–6(–8) mm. long racemosely borne along the primaries; pedicels up to about 36 per peduncle, often less, 2–11 mm. long; floral bracts obsolete. Fruits vinaceous, urceolate to subspherical, up to 5 mm. long and across, sulcate, glabrous or puberulous; stylopodium 0·5–0·7 mm. long; styles 5–8, free for 0·6–1·0 mm. terminally.

UGANDA. Karamoja District: Napak, 28 May 1940 (fr.), *A. S. Thomas* 3639! & June 1950 (fr.), *Eggeling* 5925!; Elgon, Butandiga, Feb. 1940 (fl. & fr.), *St. Clair-Thompson* in *Eggeling* 3952!
KENYA. Ravine District: Eldama Ravine, *G. A. Hoult* in *F.D.* 1092!; N. Kavirondo District: S. Elgon, June 1933 (young fr.), *Dale* in *F.D.* 3097!; Kisumu–Londiani District: Tinderet Forest Reserve, 7 July 1949 (young fr.), *Maas Geesteranus* 5385!
TANGANYIKA. Mpanda District: Mahali Mts., Wamusuku, 24 Aug. 1958 (fl.), *Newbould & Jefford* 1765! & below Kungwe Mt., Ntali R., 9 Sept. 1959 (fl.), *Harley* 9581!
DISTR. **U**1, 3; **K**3, 5; **T**4; Cameroun, Congo and Sudan Republics, Ethiopia; also Malawi and Zambia
HAB. Upland rain-forest; 1840–2770 m.

SYN. *Aralia abyssinica* A. Rich., Tent. Fl. Abyss. 1: 336 (1847)
Sciadophyllum abyssinicum (A. Rich.) Seem. in J.B. 3: 267 (1865)
Heptapleurum abyssinicum (A. Rich.) Vatke in Linnaea 40: 191 (1876); Hiern in F.T.A. 3: 29 (1877)
Schefflera acutifoliolata De Wild. in Rev. Zool. Afr. 8, Suppl. Bot.: 14 (1920) & Pl. Bequaert. 1: 155 (1921). Type: Congo Republic, Ruwenzori, valley of R. Lamya [Lamia], *Bequaert* 4310 (BR, holo.)

VARIATION. The variation in the outline of the leaflets is most striking, but since the extreme forms are not sufficiently well defined from the typical, subspecific recognition is not advisable.

3. **S. polysciadia** *Harms* in P.O.A. C: 297 (1895) & in E.J. 26: 244 (1899) & in Z.A.E.: 591 (1913); V.E. 3 (2): 779 (1921); Lebrun in B.J.B.B. 13: 19 (1934); T.S.K.: 115 (1936); F.P.N.A. 1: 690 (1948); I.T.U., ed. 2: 36 (1952); K.T.S.: 56 (1961). Types: Congo Republic/Uganda, Ruwenzori, *Stuhlmann* 2342 (B, syn. †) & Tanganyika, Kilimanjaro at Mawenzi above Kilema near Himo, *Volkens* 1877 (B, syn. †, K, isosyn. !)

Liane growing into the tree tops or a tree to 16 m. tall, woody shrub or low straggling tree. Petioles up to 22 cm. long by 6 mm. diameter, glabrous; leaflets (4–)5–7(–8), coriaceous, narrowly elliptic to rotund, lanceolate to ovate or oblanceolate to narrowly obovate, sometimes somewhat oblong, up to 28 cm. long by 12·5 cm. wide, mucronate to long-acuminate, rounded to broadly cuneate, often subcordate, occasionally ± truncate at the base, with subentire to repand, sometimes very slightly crisped margins, thickened or unthickened at extreme periphery, glabrous, with many very closely spaced parallel lateral veins; petiolules 1·0–8·5 cm. long by 1–3 mm. diameter. Stipules up to (and probably exceeding) 4·5 cm. long, adnate to petiole for up to 1·5 cm. Inflorescence (fig. 5/9) a group of up to about 6 primary branches (panicles of umbellules) 2·5–14·0(–18·0) cm. long, each typically with a racemose arrangement of secondary branches for some distance and an aggregation (umbel) at the apex; secondary branches (0·8–)2·0–7·0 cm. long with tertiary branches (peduncles of umbellules) aggregated at their ends (occasionally also with a few scattered along their length); peduncles of umbellules 0·75–3·2 cm. long; pedicels up to 8(–11) mm. long, up to about 12 per peduncle; bracts inconsistently present with large variation in size in those subtending first and second order branches, those subtending tertiary branches or pedicels very small or obsolete. Fruits urceolate to subspherical, up to 4·5 mm. long by 5·0 mm. diameter, deeply sulcate when mature, glabrous or puberulous; stylopodium 0·5–2·0 mm. long; styles 5 (rarely 6), free for up to 1·2 mm. terminally.

UGANDA. Ruwenzori, Nyinabitaba Ridge, July 1938 (fr.), *Eggeling* 3786 ! & Nyinabitaba, *Fishlock & Hancock* 196 ! & without precise locality, May 1894 (fr.), *Scott Elliot* 7785 !
KENYA. SE. Aberdare Mts., 29 July 1913 (fl. & young fr.), *Moon* 764 !; Teita Hills, Yale Peak, 13 Sept. 1953 (fl.), *Drummond & Hemsley* 4323 ! & Teita Hills, Mbololo Hill, *Gardner* in *F.D.* 2931 !
TANGANYIKA. SW. Kilimanjaro, Feb. 1928 (fl.), *Haarer* 1172 !; W. Usambara Mts., Shagai Forest, Shagayu Peak, 15 May 1953 (fl. & fr.), *Drummond & Hemsley* 2534 !; Kondoa District: Kinyassi Scarp, 2 Jan. 1928 (fl.), *B. D. Burtt* 954 !
DISTR. **U**2; **K** ?3, 4, ?6, 7; **T**2–7; Congo Republic, Ethiopia, Malawi
HAB. Moist bamboo thickets, upland rain-forest; 1540–2770 m.

SYN. *S. bequaertii* De Wild. in Rev. Zool. Afr. 8, Suppl. Bot.: 11 (1920) & Pl. Bequaert. 1: 157 (1921). Type: Congo Republic, Ruwenzori, Lanuri, *Bequaert* 4405 (BR, holo.)
S. congesta De Wild. in Rev. Zool. Afr. 8, Suppl. Bot.: 12 (1920) & Pl. Bequaert. 1: 158 (1921). Type: Congo Republic, Ruwenzori, *Bequaert* 3721 (BR, holo.)
S. angiensis De Wild. in Rev. Zool. Afr. 9, Suppl. Bot.: 35 (1921) & Pl. Bequaert. 1: 156 (1921); Chiov., Racc. Bot. Miss. Consol. Kenya: 50 (1935). Type: Congo Republic, Kivu Province, between Angi and Tongo, *Bequaert* 5855 (BR, holo.)
S. nyasensis De Wild., Pl. Bequaert. 4: 348 (1928). Type: Tanganyika, Rungwe Mt., Mbaka stream, *Stolz* 1602 (BR, holo., BM, EA, K, iso. !)
S. sp. sensu Check-lists For. Trees & Shrubs Brit. Emp. No. 1, Uganda Prot.: 24 (1935) quoad spec. *Dawe* 550 !, *Scott Elliot* 7785 !, *Fishlock & Hancock* 196 !

4. **S. volkensii** (*Engl.*) *Harms* in E. & P. Pf. 3 (8): 36, fig. 7 (1894); P.O.A. C: 297 (1895); V.E. 3 (2): 777, fig. 323 (1921); Battiscombe, Cat. Trees Kenya Col.: 86 (1926); Check-lists For. Trees & Shrubs Brit. Emp., No. 1, Uganda Prot.: 24 (1935); T.S.K.: 115 (1936); T.T.C.L.: 61 (1949); I.T.U., ed. 2: 36 (1952); K.T.S.: 56 (1961). Types: Tanganyika, Kilimanjaro, *Volkens* 986 (B, syn. †) & 1297 (B, syn. †, K, isosyn. !)

Scandent shrub or tall (sometimes spreading and much branched) tree up to 24–30 m. tall, or an epiphyte upon other trees. Petiole up to 13(–17) cm. long by 3·2(–3·8) mm. diameter, sometimes somewhat expanded at base, glabrous or sparsely hairy at tip; leaflets 4–6(–7), generally coriaceous, narrowly obovate or elliptic, occasionally oblanceolate or obovate, rarely narrowly or broadly elliptic, up to 15 cm. long by 7 cm. wide, acute to rounded or rarely slightly retuse at apex, with a generally acute, sometimes acute to rounded or broadly to narrowly cuneate base, with entire to very slightly repand, sometimes very slightly crisped margins, glabrous; petiolules up to 2·1(–2·8) cm. long by 1·9 mm. diameter. Stipules sheath-like, up to about 1·2(–1·4) cm. long. Inflorescence (fig. 5/8) a ± extended or compressed raceme of bracteate racemes of small pedunculate ± globular capitula up to 7 mm. diameter when flowers in bud; primary branches up to 23 cm. long by 4·0(–5·5) mm. diameter, generally sparsely lenticellate; secondary branches (peduncles of capitula) (5–)10(–17) mm. long, sometimes lenticellate, borne in the axils of ovate to oblate bracts up to 4 mm. long. Flowers sessile, up to 12–20 together. Fruits ± urceolate, up to 5·5 mm. long by 4(–5) mm. diameter, ± 5-ribbed, minutely puberulous or glabrous; stylopodium up to 7 mm. long; styles 5, free and spreading for up to 0·5 mm. terminally.

UGANDA. Elgon, 27 Oct. 1923 (fl. & young fr.), *Snowden* 805! & Jan. 1918 (young fr.), *Dummer* 3605! & Feb. 1940 (fr.), *St. Clair-Thompson* in *Eggeling* 3958!
KENYA. Trans-Nzoia District: Elgon, eastern slope above Tweedie's saw-mill, 25 Feb. 1948 (young fr.), *Hedberg* 162!; Nakuru District: eastern Mau Forest Reserve, 30 Aug. 1949 (fr.), *Maas Geesteranus* 5975!; Machakos/Masai Districts: Chyulu Hills, 24 Apr. 1938 (young fr.), *Bally* 340, 470, 1151 in *C.M.* 7963!
TANGANYIKA. Masai District: Ketumbaine Mt., 9 Jan. 1936 (fl.), *Greenway* 4293!; Arusha District: Meru Forest Reserve, Narok, Dec. 1951 (fl.), *Watkins* 577!; SW. Kilimanjaro, Feb. 1928 (fl. & young fr.), *Haarer* 1176!
DISTR. **U**3; **K**3–6; **T**2; Ethiopia
HAB. Upland rain-forest, upland dry evergreen forest; 1600–3230 m.

SYN. *Heptapleurum volkensii* Engl. in E.J. 19, Beiblatt 47: 41 (1894)

5. **S. stolzii** *Harms* in E.J. 53: 358 (1915); V.E. 3 (2): 778 (1921); T.T.C.L.: 61 (1949). Type: Tanganyika, Rungwe Mt., *Stolz* 2044 (B, holo. †, BM, K, iso.!)

Liane with grey-brown bark and stems up to 20 m. long and as thick as an arm (*fide* Harms). Petiole up to 22 cm. long by 3·5 mm. diameter, glabrous; leaflets 5–6, coriaceous, narrowly ovate or ovate to narrowly obovate or obovate, sometimes elliptic or broadly elliptic, up to 15 cm. long by 8 cm. wide, long-acuminate, with a rounded to acute, occasionally subcordate, rarely oblique base, with repand to almost entire, sometimes very slightly crisped margins, glabrous; petiolules 2·5–6 cm. long by 1·5 mm. diameter. Stipules sheath-like, up to about 7 mm. long. Inflorescence (fig. 5/8) an umbel of bracteate racemes of small shortly pedunculate ± globular capitula 5–7 mm. diameter when flowers in bud; primary branches up to 25 cm. long by 3 mm. diameter, with scattered lenticels; secondary branches (peduncles of capitula) up to 5(–9) mm. long, borne in the axils of puberulous, broadly ovate to deltoid bracts up to almost 3 mm. long. Flowers sessile, up to about 10–12 together. Fruits ± urceolate, 3–4 mm. long by 3–4 mm. diameter, ± 5-ribbed, minutely puberulous or glabrous; stylopodium up to 7 mm. long; styles 5, free and spreading for up to 0·5 mm. terminally.

TANGANYIKA. Rungwe District: W. slopes of Rungwe Mt., 11 Mar. 1932 (fl. & young fr.), *St. Clair-Thompson* 866! & Kiwira Forest Reserve, 8 Aug. 1962 (young fr.), *Mgaza* 466!
DISTR. **T**7; not known elsewhere
HAB. Upland rain-forest; 1600–about 2100 m.

INDEX TO ARALIACEAE

GEOGRAPHICAL DIVISIONS OF THE FLORA

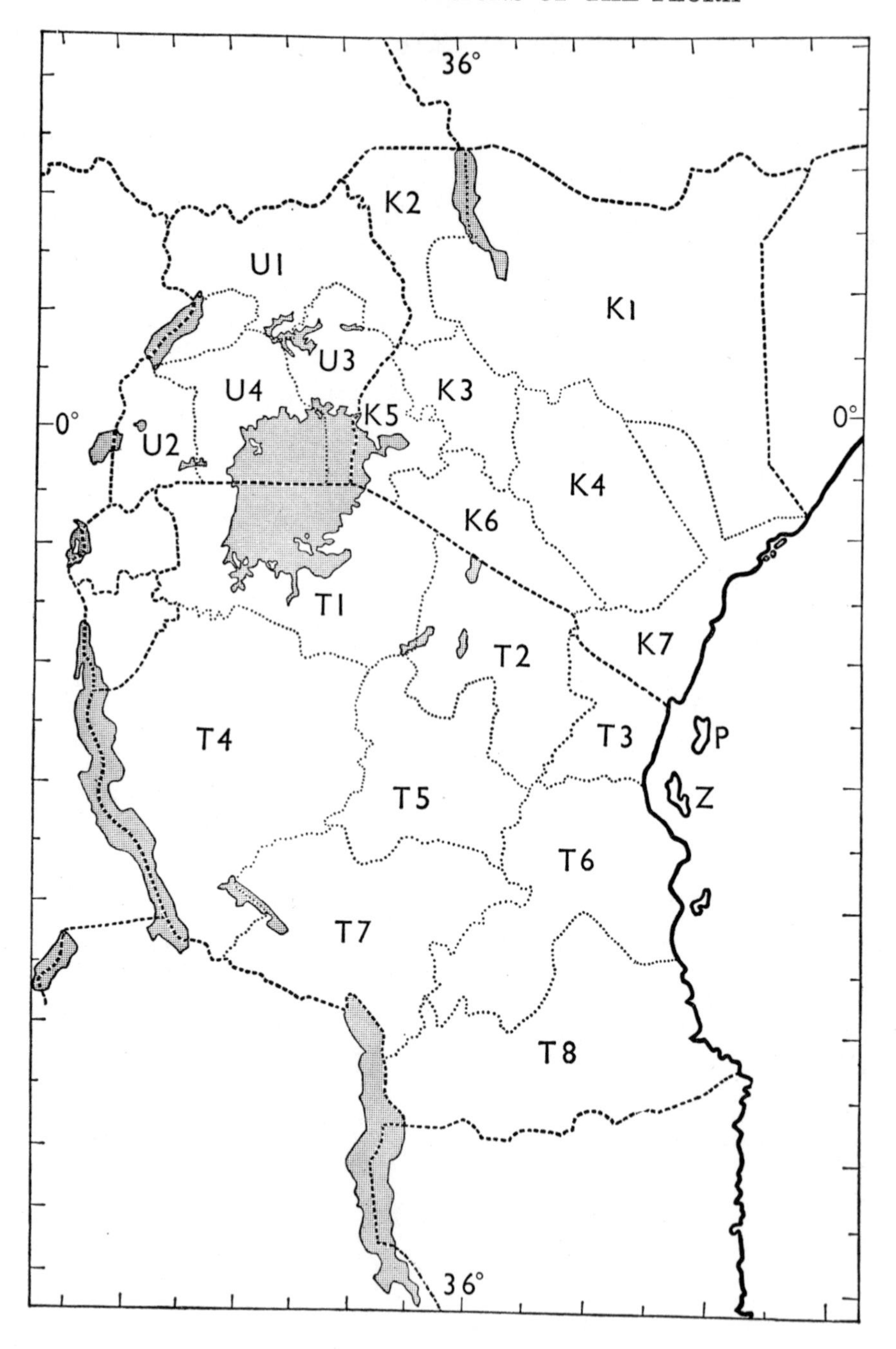